TREASURES OF

THE NATIONAL
AIR AND SPACE
MUSEUM

TREASURES OF

THE NATIONAL
AIR AND SPACE
MUSEUM

Martin Harwit and the Staff of
the National Air and Space Museum,
Smithsonian Institution

A TINY FOLIO™
ABBEVILLE PRESS PUBLISHERS
NEW YORK LONDON PARIS

FRONT COVER: "SPIRIT OF ST. LOUIS" (see pages 58 and 59).
BACK COVER: APOLLO LUNAR MODULE (TEST VEHICLE), 1969 (see page 178).
SPINE: MERCURY SPACESUIT (see page 220).
PAGE 1: CURTISS F9C-2 SPARROWHAWK (see page 83).
PAGE 2: VIKING ROCKET, C. 1955 (CENTER). Developed as a replacement for
the aging World War II V-2s (LEFT) confiscated and used for research in America.
PAGE 3: APOLLO 11 COMMAND MODULE "COLUMBIA," 1969 (see page 225).
PAGE 6: Alejandro Otero, Venezuelan. "DELTA SOLAR," 1976. Nickel and
stainless steel, 27 ft. × 46 ft. 8 ½ in. (823 × 1424 cm).
PAGE 10: NATIONAL AIR AND SPACE MUSEUM, SMITHSONIAN INSTITUTION.
PAGE 20: NOSE OF THE "SPIRIT OF ST. LOUIS" (see pages 58 and 59).

First edition
10 9 8 7 6 5 4 3 2 1

LIBRARY OF CONGRESS CATALOGING-IN-PUBLICATION DATA
National Air and Space Museum.
    Treasures of the National Air and Space Museum / Martin Harwit and
the staff of the National Air and Space Museum.
        p.   cm.
    Includes index.
    ISBN 1-55859-822-7
    1. National Air and Space Museum—Catalogs.   2. Aeronautics—United
States—History.   3. Astronautics—United States—History.   I. Harwit,
Martin, 1931– . II. Title.
TL506.U6W376   1995
629.1'074'753—dc20   94-41496

CONTENTS

———

FOREWORD

——————

I T IS ALMOST guaranteed that every visitor who walks through the doors of the National Air and Space Museum for the first time will pause and look up. For right there, suspended in the airy central space of the Milestones of Flight gallery, are the original Wright 1903 Flyer (the first true airplane), the *Spirit of St. Louis* (which made the first solo transatlantic flight), the Bell X-1 (the first supersonic airplane), and the X-15 (the world's fastest airplane). Below them stands the Apollo 11 command module— the spacecraft that brought the first men to land on the Moon back to Earth. Here, in this gallery, it is only natural to experience something of the great, soaring, liberating sense of flight.

From the Wright brothers to Apollo 11 was a mere sixty-six years—less than a lifetime—but these years witnessed the most extraordinary outpouring of innovation and achievement in the technology and art of aviation and spaceflight, with consequences that changed the world forever.

Extending outward from the Milestones of Flight gallery are other galleries from which more wonders beckon. The airplanes in the collection represent the best of their time in aviation design and performance, their beautiful forms reflecting our evolving mastery of aerodynamics. Some of these aircraft can be appreciated even as works of art. The rockets and spacecraft trace the story of how our species has finally unlocked the shackles of Earth's gravity, allowing us to see and study our planet from a different viewpoint, and to launch robot voyages to distant worlds.

This little book beautifully depicts the highlights of the finest collection of airplanes and spacecraft in the world. We hope you will enjoy it.

MARTIN HARWIT
*Director, National Air
and Space Museum*

INTRODUCTION

O**N A WINDY** December day in 1903, Orville and Wilbur Wright took turns at the controls of a fragile-looking biplane and flew above the beach at Kitty Hawk, North Carolina. Though the longest of their flights lasted a mere fifty-nine seconds and skimmed just above the sand for 852 feet (260 m), on that historic day the Wright brothers realized a dream that had enthralled the adventurous for centuries. They became the first humans in history to leave the Earth on board a successful, controlled, powered, heavier-than-air machine.

Buoyed by the Wright brothers' success, succeeding generations of aviation pioneers have been relentless in their quest to fly. In 1927, a quarter of a century after the Wright brothers flew across the beach at Kitty Hawk, Charles A. Lindbergh flew nonstop across the Atlantic Ocean; and thirty-five years after that, in 1962, John Glenn orbited the Earth in a spacecraft. Today, not even a full century after that day in Kitty Hawk, manned flight has put the farthest reaches of our own planet and nearby parts of our universe within reach. We routinely race around the globe on jet-powered aircraft, send explorers to neighboring

worlds on missions that the Wright brothers could never have imagined, and enjoy the benefits of global communication made possible by the vast network of satellites that continually orbit the Earth.

Many of the most exciting and significant artifacts of the history of flight—nearly 370 aircraft, as well as numerous rockets, missiles, and spacecraft —are preserved in the collections of the National Air and Space Museum. The museum is part of the Smithsonian Institution and shares the Institution's mandate to increase and diffuse knowledge; the specific mission of the National Air and Space Museum is to examine and present the history, science, technology, and social impact of aeronautics and spaceflight and to investigate and exhibit the nature of the universe and the environment.

The collections that now compose the National Air and Space Museum took root in 1876 when, at the close of the Philadelphia Centennial Exposition, the Chinese Imperial Commission presented the Smithsonian Institution with a gift of kites. Ever since this modest beginning, the Smithsonian Institution had shown interest in flight. Samuel P. Langley, the Institution's third Secretary, was himself a pioneer in aviation, and in 1896 successfully launched an unmanned flying machine off the top of a houseboat on the Potomac River. The small craft covered more than half a mile. His attempt at piloted flight, however, met a wet and disappointing end. A full-size airplane based on the earlier successful model crashed into the Potomac upon takeoff in 1903. Langley's successor, Charles D. Walcott, played a major role in creating the National Advisory Committee for Aeronautics (NACA), established in 1915. NACA was the forerunner of today's National Aeronautics and Space Administration (NASA). The Smithsonian entered space research early in the game; beginning in 1916 it supported the rocketry experiments of Robert H. Goddard.

In 1946, an act of Congress mandated that the Smithsonian Institution collect and exhibit material memorializing "the national development of aviation" in a National Air Museum. In 1966, the museum expanded its charter to include spaceflight and has since been known as the National Air and Space Museum. The Smithsonian's interest in flight reflects the enthusiasm of the public: since the museum opened its current building to the public on July 1, 1976, an average of more than eight million have visited its galleries each year.

The Wright 1903 Flyer (page 28) is the centerpiece of the museum's Milestones of Flight gallery. It is suspended in mock flight above and to the side of Gemini IV (page 218), the spacecraft from which astronaut Edward H. White II emerged on June 3, 1965, to float high above North America, becoming the first American to walk in space. The *Spirit of St. Louis* (pages 20, 58, 59), the airplane in which Charles Lindbergh flew from New York to Paris, proving that an airplane could cross vast stretches of the Earth, hangs in a gallery not far from a Douglas DC-3, a luxury liner that took flight in 1935. The DC-3 could carry fourteen passengers in sleeping berths or twenty-one in seats. This venerable airplane spanned the modern era of commercial passenger transport.

The Wright EX *Vin Fiz* (page 31), looking fragile and tiny amid the larger aircraft in the Pioneers of Flight gallery, is the first airplane to have crossed the United States. Cigar-smoking pilot Calbraith Perry Rodgers crashed nineteen times, made sixty-nine stops, and used enough spare parts to build four new planes. The trip took forty-nine days, with a total flying time of eighty-two hours and two minutes. Performance had improved considerably by August of 1932, when Amelia Earhart stepped into the cockpit of a shiny, bright red Lockheed Vega 5B (page 69) and flew from Los Angeles to Newark, New Jersey, in nineteen hours and five minutes, the first nonstop transcontinental flight by a woman. In May of the same year, Earhart had set off from Harbor Grace, Newfoundland, also alone at the controls, to become

The camera records an historic flight on December 17, 1903, at Kitty Hawk, North Carolina. As his brother Wilbur watches, Orville Wright pilots the Flyer in the first successful powered flight.

the first woman to make a nonstop solo flight across the Atlantic.

The world's first military airplane was built by the Wright brothers and sold to the U.S. Army in 1909. The brothers demonstrated the Wright Military Flyer (page 30) for President Taft and military brass at Fort Meyer, Virginia. Among the museum's collection of other military aircraft is the Bell XP-59A (page 106). It was designed amidst great secrecy and, in 1942, was the first American jet airplane to fly. High-speed flight technology zoomed ahead quickly. In 1947, Air Force test pilot Captain Charles "Chuck" Yeager climbed into the cockpit of the Bell X-1 *Glamorous Glennis* (page 131), named for his wife. At an altitude of forty-three thousand feet over the Mojave Desert, traveling faster than the speed of sound for the first time and ushering in the age of supersonic flight, Yeager reached Mach 1.06, approximately seven hundred miles per hour. In addition to such notable aircraft, the museum's aeronautical holdings include engines, propellers, instruments and avionics, navigational aids, flight material, and awards and memorabilia, making this collection the finest of its kind. Specimens from the National Air and Space Museum's collections are routinely lent to and displayed at other institutions around the world.

The rockets, missiles, and spacecraft of the

National Air and Space Museum make up the definitive collection of artifacts from the U.S. space program. Through an agreement forged with the National Aeronautics and Space Administration (NASA) in 1967, the museum has become the primary custodian of the historic American relics of space exploration, which it displays, lends to other museums, and preserves in storage and study collections. The Space History collection includes all flown, manned, U.S. spacecraft, except Gus Grissom's Mercury capsule *Liberty Bell 7* (which was lost at sea) and an operational Space Shuttle.

A wide range of satellites, including a reproduction of Sputnik 1 (page 196)—launched by the Soviet Union in 1957 and the first artificial satellite to orbit the Earth—represent almost four decades of space exploration. Other spacecraft and related objects represent space exploration's amazing achievements. On January 31, 1958, a Jupiter-C rocket put a U.S. satellite, Explorer 1 (page 197), into orbit for the first time. On May 5, 1961, Alan B. Shepard, Jr., America's first astronaut, stepped into the Mercury spacecraft *Freedom 7* (page 216), was lifted 116 miles into suborbit from Cape Canaveral, and splashed down in the Atlantic Ocean fifteen minutes and twenty-two seconds later. American manned space flight was born.

Soon the Gemini and Apollo missions were under way to meet the goal of landing a man on the Moon and returning him safely within the decade. In December 1968, astronauts first orbited the Moon, and in July 1969 the first humans walked on its surface. By May of 1973, Americans were living in space for prolonged periods. From May 25, 1973, to February 8, 1974, the Skylab Orbital Workshop (pages 242, 244, 245) was home to three crews of astronauts, who spent, respectively, twenty-eight, fifty-nine, and eighty-four days on the space station. This program is represented in the museum by flight back-ups (Skylab orbited the Earth until July 11, 1979, when it crashed harmlessly into the Indian Ocean and the deserts of Australia). On July 17, 1975, the American Apollo spacecraft and the Soviet Soyuz joined in space (page 247), allowing crew members to visit each other's craft. The historic docking, representing a milestone in space exploration as well as in international relations, is recreated in the museum. The Space Shuttle test vehicle *Enterprise* (pages 248 and 249) represents America's most recent foray into manned spaceflight. The world's first re-usable space vehicles do yeoman's service in putting satellites and space labs into orbit. On display in the museum are models of the Hubble Space Telescope (one of many space-based observatories on display), which the Shuttle launched into orbit in 1990 and which has transmitted detailed images of our universe and neighboring planets.

The Space History collection also includes 122 spacesuits, including most of those used from Project Mercury through the Apollo-Soyuz Test Project. Among them are those worn by the Apollo astronauts who first set foot on the Moon. Unmanned planetary spacecraft, such as the Viking Mars Lander (page 234), which has increased our knowledge of Mars significantly through its work on the planet's surface, and the Voyager probes sent on a grand tour of Jupiter, Saturn, and the outer planets, are represented by backup and test models of almost all of the American robotic explorers. The flown spacecraft do not return to Earth and so cannot be collected.

In addition to its exhibits, the National Air and Space Museum performs many other services. Two historical departments, Aeronautics and Space History, conduct research on the origin and development of flight through the atmosphere and in space. Two scientific laboratories, the Center for Earth and Planetary Studies and the Laboratory for Astrophysics, carry out basic research in satellite remote sensing of the environment, planetary surfaces and atmospheres, observational and theoretical astrophysics, and the development of infrared astronomical instruments for spacecraft. An extensive library and archives are available to the public for research. The museum's Albert Einstein Planetarium is one of the most popular and technically sophisticated in the world. The Samuel P. Langley Theater presents a series of IMAX films on a five-story-high screen, as well as lectures and seminars.

The museum's Paul E. Garber Preservation, Restoration and Storage Facility in Suitland, Maryland, contains a large portion of the air- and spacecraft collection and is open for tours by reservation. Early in the next century, the museum will open an extension at the Washington Dulles International Airport, providing expanded preservation and storage facilities as well as buildings to exhibit the larger airplanes and spacecraft in the collection that cannot now be displayed.

For succeeding generations of enthusiasts, the National Air and Space Museum will continue to capture the importance of these milestones and, through growth and change, it will reflect the evolving impact of aviation and spaceflight on our world. Beginning with that blustery day at Kitty Hawk, powered flight has changed the manner in which people travel, wage war, and look at our planet and the cosmos. Few technological achievements of the twentieth century have had as profound an effect on humanity.

AVIATION

FLIGHT HAS long fascinated humankind, from the ancient Greeks, who related the myth of Daedalus and Icarus soaring into the sky, through the great Renaissance artist Leonardo da Vinci, who created remarkable sketches of flying machines. It was only in the late nineteenth century, however, that significant progress was made in the field of heavier-than-air flight. Otto Lilienthal's pioneering aerodynamic research and the gliders he built and flew from 1891 to 1896 laid the foundation for the Wright brothers' successful powered aircraft of 1903. Less than a decade after the Wright brothers' tentative flights, aircraft were being used in war. Both bombing and reconnaissance missions were carried out in 1911 over Tripoli by the Italians. Similar tactics were employed during the various Balkan wars that preceded World War I.

Charles Lindbergh made his historic solo transatlantic flight in 1927, a time when barnstorming pilots dazzled crowds with hair-raising aerobatics and commercial transportation and air mail delivery were emerging. The Ford Trimotor (pages 62–63) and Pitcairn Mailwing (page 60) were mainstay aircraft in the early period of commercial transport. Commercial aviation really took off in the 1930s, with the production of modern aircraft, starting with the Boeing 247D (page 81) and Douglas DC-3 (page 85). The 1930s are also noted for record-breaking flights by such pilots as Wiley Post and Howard Hughes.

The 1940s was one of the most significant decades in the history of aviation, with the development of jet aircraft and long-range rockets—breakthroughs that would have a profound impact in the years to follow. The Grumman F4F Wildcat (page 90) and the Mitsubishi A6M5 Zero (page 100) represent early World War II technology. The Wildcat served as the United States Navy's principal fighter at the onset of the war and was later succeeded by the F6F Hellcat (page 91) and the Vought F4U-1D Corsair (page 94). Deployed in conjunction with the TBF-1 Avenger (page 92) torpedo bomber, the Douglas SBD-6 Dauntless (page 93) served as the Navy's dive-bomber. Both saw action in every major battle in the Pacific.

Over the past forty years, technology has continued to advance and has made flight a commonplace part of life in the late twentieth century. Chuck Yeager's 1947 historic Mach 1 flight in the Bell X-1 proved that supersonic flight was feasible, and aviators have continued to push the limits in air speed and flight duration with aircraft such as the SR-71 Blackbird and the Rutan *Voyager*. The museum's aviation exhibits shed light on just how rapidly technology has developed in less than a century, and the visitor cannot help but speculate on what tomorrow may bring.

MONTGOLFIER BALLOON (MODEL)
Carried humans aloft for the first time over Paris
on November 21, 1783.

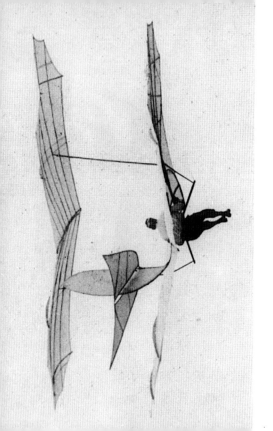

LILIENTHAL GLIDER, 1894
Many of Lilienthal's glides were made from an artificial hill constructed near his home in a Berlin suburb.

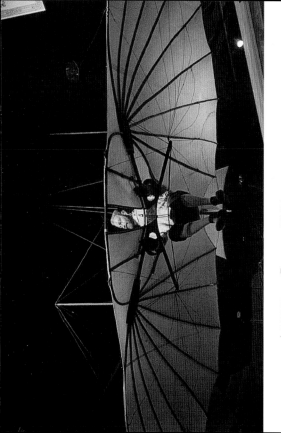

LILIENTHAL GLIDER, 1894
Lilienthal's lift experiments were an important starting point for many later aeronautical researchers, including the Wright brothers.

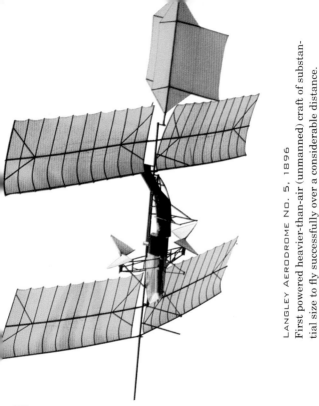

LANGLEY AERODROME NO. 5, 1896
First powered heavier-than-air (unmanned) craft of substantial size to fly successfully over a considerable distance.

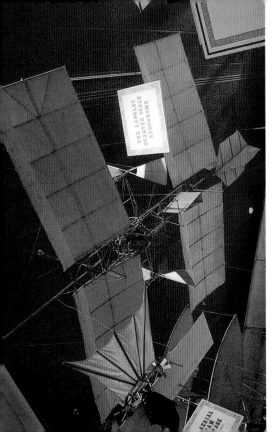

LANGLEY AERODROME, 1901
Quarter-scale model used in balance studies during the design and construction of the full-scale piloted Aerodrome A of 1903.

WRIGHT 1903 FLYER
First controlled, powered, piloted, heavier-than-air craft to fly.

ORVILLE WRIGHT

Close-up of the pilot's position in the Wright 1903 Flyer.

WRIGHT 1909 MILITARY FLYER
World's first military airplane.

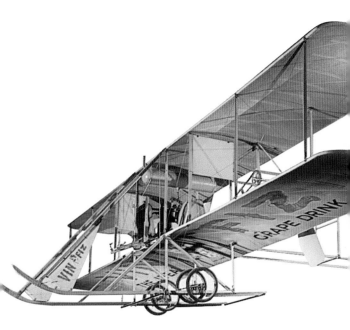

WRIGHT EX "VIN FIZ," 1911
First airplane to make a transcontinental flight across
the U.S.

CURTISS D "HEADLESS PUSHER," 1912

Standard biplane of Curtiss Exhibition Team and one of the most successful aircraft of the pioneer era.

THE CURTISS HEADLESS PUSHER

WISEMAN-COOKE BIPLANE, 1910
Made the first locally authorized
airmail flight, from Petaluma to Santa
Rosa, California, in February 1911.

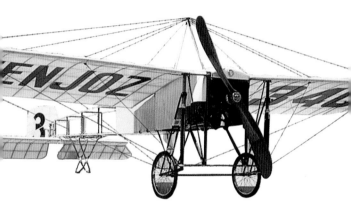

BLÉRIOT XI, 1914

Similar to the aircraft in which Louis Blériot first crossed the
English Channel on July 25, 1909.

COMPLIMENTS OF
MISS HARRIET QUIMBY

HARRIET QUIMBY

First American woman to receive her pilot's license and the second in the world. In 1912, with her Blériot XI, she became the first woman to fly the English Channel.

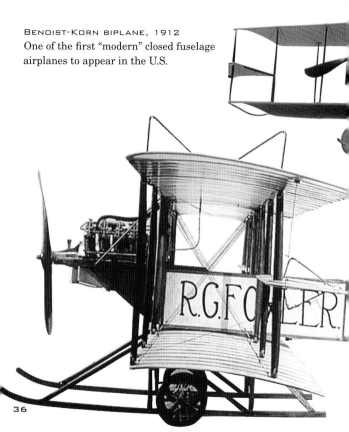

BENOIST-KORN BIPLANE, 1912
One of the first "modern" closed fuselage
airplanes to appear in the U.S.

R.G.FO LER

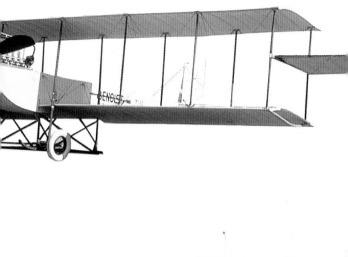

Fowler-Gage biplane, 1912

Example of the forward-engine biplanes of the pre–World War I era. It made the first aerial crossing of the Isthmus of Panama.

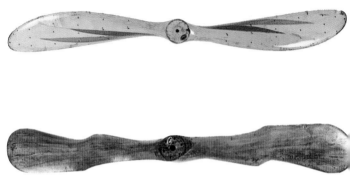

TOP. CHAUVIÈRE PROPELLER
Constructed of laminated wood, one of the finest European propellers.

BOTTOM. REQUA GIBSON PROPELLER, 1911
Manufactured by E. W. Bonson, and used by Professor David L. Gallup in experiments at Worcester Polytechnic Institute.

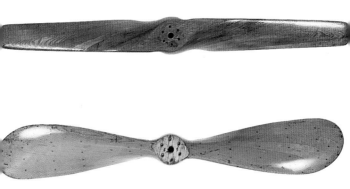

TOP. GALLUP PROPELLER, 1911
Used by Professor David L. Gallup in experiments at
Worcester Polytechnic Institute.

BOTTOM. PARAGON PROPELLER, 1911
Manufactured by the American Propeller Company of
Baltimore, and used by Professor David L. Gallup in
experiments at Worcester Polytechnic Institute.

ECKER FLYING BOAT, C. 1912

Essentially a copy of the highly successful Curtiss Flying Boat, and one of the oldest flying boats in existence.

SPAD XIII "SMITH IV," 1918

French fighter flown by many American squadrons in the last year of World War I.

Curtiss NC-4 (¹/₂₄ scale model), 1919
First aircraft to cross the Atlantic Ocean.

Sopwith Snipe, 1918
Descendant of the famous Sopwith Camel;
became the Royal Air Force's first standard
postwar fighter.

CAUDRON G.4, 1915
French twin-engined observation airplane; also
served as a bomber and a trainer.

VOISIN MODEL 8, 1916
French aircraft used for strategic bombing and
observation in World War I.

CURTISS JN-4D "JENNY," 1917
Most important American, World War I–
era trainer.

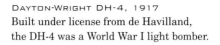

DAYTON-WRIGHT DH-4, 1917
Built under license from de Havilland,
the DH-4 was a World War I light bomber.

ALBATROS D.VA, 1917
Produced in greater numbers than any other German
World War I fighter.

DE HAVILLAND DH-4 "OLD 249," 1918

Originally a World War I military trainer; became one of
the original fleet of aircraft for the airmail division of the
U.S. Post Office.

German fighter that appeared late in World War I; also used in several well-known Hollywood World War I films made in the 1920s and 1930s.

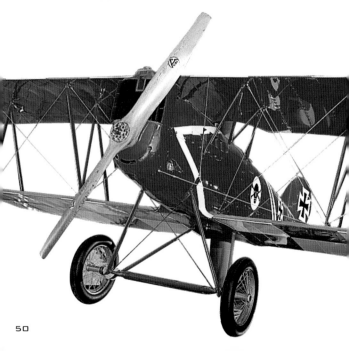

FOKKER D.VII, 1918
Introduced in 1917, and considered Germany's best
World War I fighter.

FOKKER T-2 (F.IV), 1923

Made first nonstop, coast-to-coast flight across North America in 192

DOUGLAS WORLD CRUISER "CHICAGO," 1924
First airplane to circumnavigate the world.

BELLANCA CF, 1922
High-wing monoplane with lifting struts and a fully enclosed passenger cabin.

CURTISS ROBIN J-1 "OLE MISS," 1928

Set the world record of twenty-seven days sustained flight in 1935.

Captured Schneider Cup for
seaplanes and set a world speed
record of 247.7 miles per hour.

ARMY LIEUTENANT JAMES H. DOOLITTLE stands on the float of a Curtiss R3C-2, similar to the one he piloted to victory in the 1925 Schneider Trophy Race.

RYAN NYP "SPIRIT OF ST. LOUIS," 1927
Flags on the nose represent the countries Lindbergh visited
on his tour of Latin America.

RYAN NYP "SPIRIT OF ST. LOUIS," 1927
The airplane in which Charles Lindbergh made his historic solo, nonstop, transatlantic flight from New York to Paris.

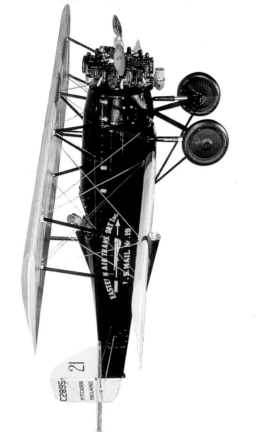

PITCAIRN PA-5 MAILWING, 1927
First airplane flown by what would become Eastern Airlines.

DOUGLAS M-2 MAILPLANE, 1926
One of the Western Air Express airplanes that flew between
Los Angeles and Salt Lake City from 1926 to 1930.

61

Aeronca C-2, 1929
Designed as an affordable airplane for private use.

FORD 5-AT TRIMOTOR, 1926

Known affectionately as the "tin goose," and among the first successful passenger-carrying aircraft.

VERVILLE SPORT TRAINER, 1931
Two-seat, tandem biplane, designed to appeal to the
wealthy private owner.

WACO 9, 1925
Popular biplane that served
as a trainer, crop duster,
and barnstormer.

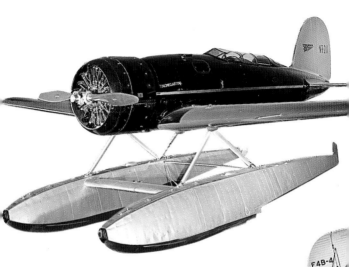

LOCKHEED SIRIUS "TINGMISSARTOQ," 1931
Used by Charles and Anne Morrow Lindbergh on
their famous transoceanic flights in 1931 and 1933.
They pioneered routes still used today for
commercial air traffic.

BOEING F4B-4, 1928

Popular biplane fighter that flew for the
U.S. Navy in the early 1930s.

NORTHROP ALPHA (TWA), 1930
Passenger transport featuring many aerodynamic and structural advancements.

LOCKHEED VEGA 5B, 1932

Aircraft piloted by Amelia Earhart on her solo, nonstop, transatlantic flight, the first by a woman, May 20, 1932.

LOCKHEED VEGA 5C "WINNIE MAE," 1931

Aircraft piloted by the famous aviator Wiley Post around the world and on high-altitude record flights.

NORTHROP GAMMA "POLAR STAR," 1935

First aircraft to complete a transantarctic flight.

BOWLUS ALBATROSS 1 "FALCON," 1930

In 1934 Warren Eaton flew this glider to 9,094 feet, an American altitude record for motorless aircraft.

PITCAIRN AC-35, 1935
Roadable autogiro that operated as both an aircraft and
an automobile.

ARLENE DAVIS
The racing pilot and official of the National Aeronautics
Association poses with her German shepherd in front of her
B17L Staggerwing, November 1935.

BEECHCRAFT C17L STAGGERWING, 1936
One of the classic designs of the Golden Age of Flight,
this corporate and racing airplane was Beechcraft's first
production aircraft.

HERRICK HV2A CONVERTOPLANE, 1936

Experimental aircraft; bridged the gap between autogiros and helicopters.

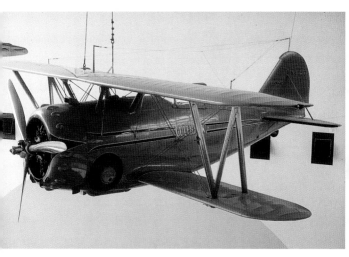

GRUMMAN G.22 "GULFHAWK II," 1936

Famous aerobatic demonstrator designed specifically for
Al Williams and the Gulf Oil Company.

HUGHES H-1 RACER, 1935

High-speed aircraft flown by mogul Howard Hughes on record-breaking flights in 1935 and 1937.

WITTMAN "BUSTER," 1931
Steve Wittman's midget racer, it competed successfully from
1931 until 1954.

Grumman G-21 Goose,
1937
Twin-engined amphibian; first
Grumman aircraft to enter
commercial service.

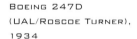

Boeing 247D
(UAL/Roscoe Turner),
1934
Commonly referred to as
the first modern airliner,
because of its all-metal
construction, cowled multi-
engines, and retractable
landing gear.

Only eight ever built, their primary function was
defense of the 1930s U.S. dirigible fleet.

Boeing P-26A Peashooter, 1932
First U.S. military airplane to feature an all-metal
monoplane construction.

BEECHCRAFT D.18, 1937

Highly successful six- to eight-passenger aircraft, with a production run of thirty-two years.

84

Douglas DC-3 (EAL), c. 1936

Highly successful transport; served in both military and commercial capacities.

PIPER J-3 CUB, 1938
Arguably the most famous light airplane in history; filled a
variety of roles over the years.

STINSON SR-10F RELIANT, 1938
First "pick-up" of a human being by an aircraft in full flight, on September 5, 1943.

WORLD WAR II AND THE
DAWNING OF A NEW ERA
(1940–1949)

Curtiss P-40E Warhawk, c. 1942
Later-model version of the famous U.S. fighter that served
with the Flying Tigers in China.

GRUMMAN F4F-4 (FM-1) WILDCAT, c. 1942

Served as the U.S. Navy's principal fighter at the beginning of World War II.

GRUMMAN F6F-3 HELLCAT, 1943
Carrier-based successor to the Wildcat; established a
19-to-1 kill-to-loss ratio in the Pacific.

DOUGLAS SBD-6 DAUNTLESS, C. 1944
Final version of the U.S. Navy's primary World War II
dive bomber.

GRUMMAN TBF-1 AVENGER, C. 1943
U.S. Navy's main torpedo bomber during World War II.

Vought F4U-1D Corsair, c. 1943 (RIGHT)

Served with the U.S. Navy and Marine Corps during World War II and the Korean War; its gull wings were necessitated by its large propeller.

Northrop N-1M Flying Wing, 1940 (BELOW)

The first flying wing airplane with pilot, engine, and fuselage integrated in a basic airfoil envelope.

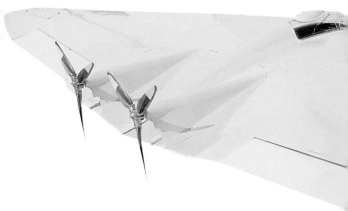

North American B-25J (TB-25M)
Mitchell, c. 1943
Later model version of the U.S. medium
bomber used in the bombing of Tokyo in 1942
by Doolittle's Raiders.

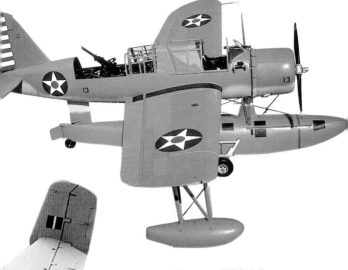

**Vought OS2U-3
Kingfisher, c. 1942**
U.S. Navy floatplane used for
rescuing downed pilots and
for reconnaissance.

BOEING B-29 SUPERFORTRESS "ENOLA GAY," 1945
Dropped the first atomic bomb on Hiroshima, Japan, on
August 6, 1945.

"ENOLA GAY" AND CREW, 1945

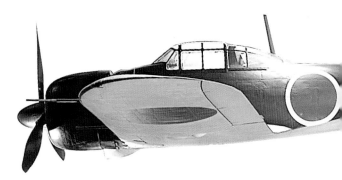

Mitsubishi A6M5 Zero,
c. 1943

Final operational version
of the primary Japanese
Navy's World War II
fighter; served in China
and the Pacific.

NAKAJIMA J1N1-S GEKKO
"IRVING," c. 1943
Japanese Navy's nightfighter.

de Havilland D.H. 98 Mosquito, c. 1944

British twin-engined nightfighter, photo reconnaissance aircraft, fighter-bomber, and target tow aircraft.

SUPERMARINE SPITFIRE
MK. VII, c. 1944
British fighter produced
throughout the war;
Museum's exhibit represents
a special high-altitude,
pressurized model.

HAWKER HURRICANE MK. IIC, C. 1942
British fighter used as a night intruder and fighter-bomber.

BELL P-39Q AIRACOBRA, C. 1942
A unique midengined fighter; many sent to the Soviet Union under the Lend-Lease Program.

Bell XP-59A Airacomet, c. 1942 (TOP)
First American jet aircraft; guide for future U.S. advances
in jet aviation.

BOEING B-17G FLYING FORTRESS, c. 1943
(BOTTOM)

U.S. long-range heavy bomber; served as the backbone of the U.S. bomber force in the strategic bombing of Germany.

48381

MARTIN B-26B MARAUDER "FLAK BAIT" (NOSE)

THE CREW OF THE 200TH MISSION OF THE
MARTIN B-26B "FLAK BAIT"

North American SNJ-4, c. 1941

Served as an advanced trainer for Navy fighter pilots throughout the 1940s; also used by the Army Air Forces as the AT-6.

NORTH AMERICAN P-51D MUSTANG, 1944
U.S. long-range escort fighter; served in both theaters of
World War II and in Korea.

NORTHROP P-61C BLACK WIDOW,
c. 1944
Twin-engined U.S. nightfighter; served
in both theaters of World War II.

REPUBLIC P-47D THUNDERBOLT,
c. 1944
Nicknamed the "JUG," the heaviest
single-engined fighter of World War II;
produced in greater numbers than any
other U.S. fighter.

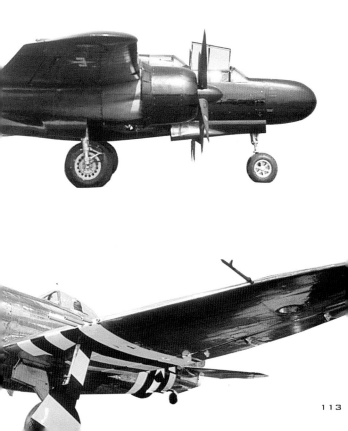

ARADO AR 234B BLITZ, 1944
German twin-engined jet aircraft; the world's
first operational jet bomber.

Arado Ar 234B Blitz (nose)

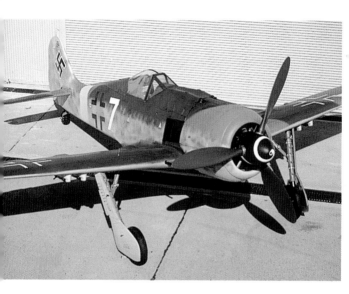

FOCKE-WULF FW 190F-8, C. 1944
One of Germany's most adaptable fighters, it served in a
variety of roles ranging from interceptor to fighter-bomber.

FOCKE-WULF FW 190F-8 (COCKPIT)

117

HEINKEL HE 162A-2 VOLKSJAGER, 1945

German jet aircraft born out of desperation and intended to be flown by the Hitler Youth; saw little combat.

DORNIER DO 335 PFEIL, C. 1944

Experimental "push/pull" design claimed by the Germans to be the fastest propeller-driven aircraft of World War II; never used operationally.

MESSERSCHMITT BF 109G-6 "GUSTAV," C. 1943

German fighter produced in greater numbers (more than thirty thousand) than any other German aircraft.

MESSERSCHMITT BF 109G-6 (COCKPIT)

MESSERSCHMITT ME 262-1A SCHWALBE, 1944
The world's first operational jet fighter to see combat.

MESSERSCHMITT ME 163B KOMET, C. 1944
German rocket-powered interceptor deployed in limited
numbers; the world's first and only operational rocket fighter.

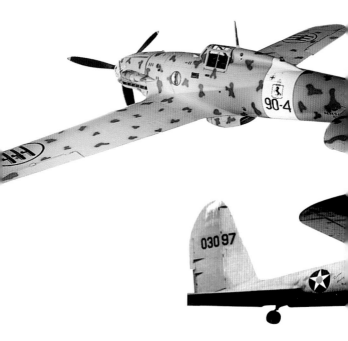

Macchi MC 202 Folgore, 1942

Advanced Italian fighter; served both in Europe and
North Africa.

BOEING B-17D FLYING FORTRESS "SWOOSE," 1941

Early model of one of the most important U.S. heavy bombers of World War II.

LOCKHEED XP-80, SHOOTING STAR, 1944
Experimental prototype of the first operational U.S. jet fighter.

Curtiss XP-55 Ascender, 1943
Experimental aircraft featuring a
pusher-type propeller, no empennage,
and fixed forward surfaces.

MCDONNELL FH-1 PHANTOM I, 1946
The first U.S. operational carrier-based jet fighter.

DOUGLAS XB-42A, 1945
Experimental bomber once
intended to supplement the
Boeing B-29; it was tested in
both jet- and propeller-
driven models.

Grumman F8F-2 Bearcat "Conquest 1,"
1946 (1969)

Aircraft in which Darryl Greenamyer broke
the world absolute speed record for piston-
engined aircraft set by Fritz Wendel in 1939.

Aircraft in which Captain Charles "Chuck" Yeager became the first person to fly faster than the speed of sound.

Focke-Achgelis Fa 330-1A Bachstelze, 1942

This unpowered autogiro was towed aloft by German
U-boats and used as an observation platform.

PENTICOST E III
HOPPICOPTER, C. 1946
Short-lived experiment
intended as an inexpensive
option for the aerial
deployment of infantry.

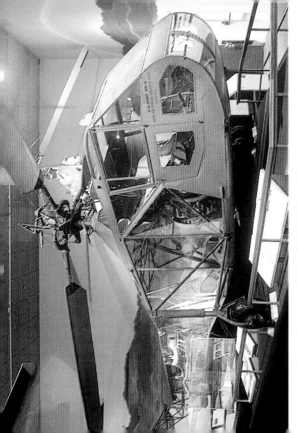

SIKORSKY XR-4 (VS-316), 1942
First mass-produced helicopter in the world.

PIASECKI PV-2, 1943
Second helicopter to fly successfully in the U.S.

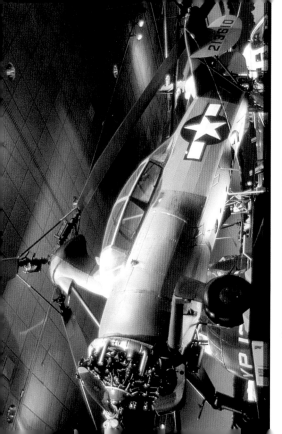

Kellett XO-60 Autogiro, 1943

Autogiro tested by the U.S. Army for an observation craft; never reached production stage.

HILLER XH-44 HILLER-COPTER, 1944
Innovative experimental helicopter design; featured twin coaxial counter-rotating blades and no tail rotor.

NX30033

Hiller-copter

SIKORSKY XR-5 DRAGONFLY, 1943

Flying prototype of the first helicopter Sikorsky sold on the commercial market after World War II.

BEECHCRAFT 35 BONANZA, 1947

Set record for light plane flight from Hawaii to the continental
U.S., and from Honolulu to Teterboro, New Jersey, both in 1949.

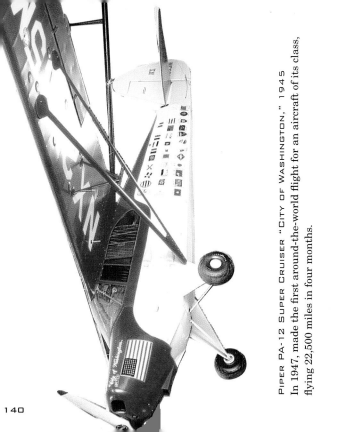

PIPER PA-12 SUPER CRUISER "CITY OF WASHINGTON," 1945
In 1947, made the first around-the-world flight for an aircraft of its class, flying 22,500 miles in four months.

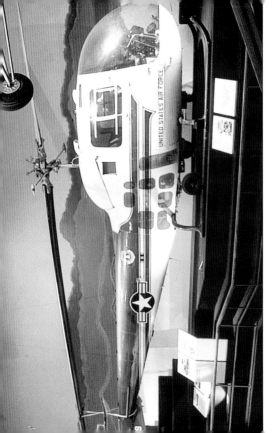

BELL VJ.13J RANGER, 1945
Dwight D. Eisenhower became the first President of the U.S. to travel by helicopter in this aircraft on July 12, 1957.

NORTH AMERICAN
P-51C EXCALIBUR III,
C. 1942 (RIGHT)
Flown by Charles Blair on
a historic solo flight over
the North Pole in 1951.

DOUGLAS A-1H SKYRAIDER,
c. 1945 (ABOVE)
Rugged attack aircraft that
served with the U.S. Navy,
Marines, and Air Force during
the Korean and Vietnam wars.

PIPER PA-18 SUPER CUB,
1955

The popular post–World War
II, two-seat cabin monoplane
was used for a wide variety of
private and utility flying.

CESSNA 180 "SPIRIT OF COLUMBUS," 1953

Popular four-seat light airplane; in 1964, first aircraft piloted by a woman to fly around the world.

BELL ATV VERTICAL TAKEOFF AND LANDING (VTOL), 1954

Test aircraft developed quickly and inexpensively using Schweizer glider fuselage, Cessna 150 wing, helicopter landing gear, and motor boat throttle.

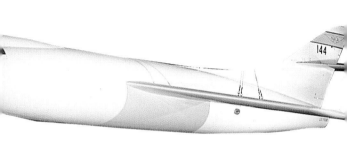

Douglas D-558-2, 1953
Aircraft in which Scott Crossfield became the first person to fly at twice the speed of sound.

NORTH AMERICAN F-86A SABRE, 1950
Developed in the late 1940s; best remembered for success against Soviet jet aircraft during the Korean War.

DOUGLAS DC-7, C. 1953

Four-engined long-range passenger transport; popular in the 1950s.

CONVAIR XFV-1 "POGO," 1954

Designed to take off and land vertically; "POGO" was tethered inside a hangar during initial flights.

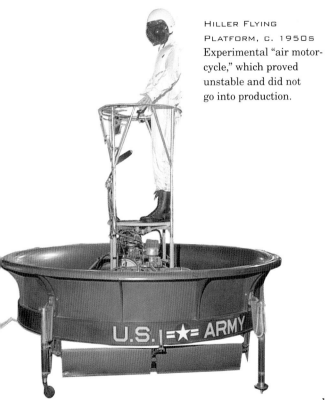

HILLER FLYING
PLATFORM, c. 1950s
Experimental "air motor-
cycle," which proved
unstable and did not
go into production.

U.S. ★ ARMY

BOEING 367-80 (707), 1954

Prototype of first successful U.S. commercial air transport aircraft; milestone in transition from piston- to jet-powered passenger aircraft.

Douglas A-4C (A4D-2N)
Skyhawk, 1954
Very successful light bomber; served in Vietnam, with the Blue Angels, and in air forces around the world.

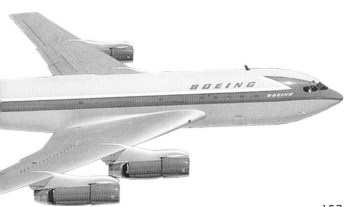

BENSEN B-6 GYROGLIDER, 1954
Home-built autogiro towed aloft by car or boat.

Powered version of the gyroglider, it set twelve world and national autogiro speed, distance, and altitude records between May 1967 and June 1968.

SIKORSKY UH-34D (S-58) CHOCTAW, 1954

Helicopter that served in a variety of roles from medivac to troop transport with the U.S. Army and Marines.

LOCKHEED U-2C, 1955
America's premier high-altitude reconnaissance airplane during the 1950s and 1960s.

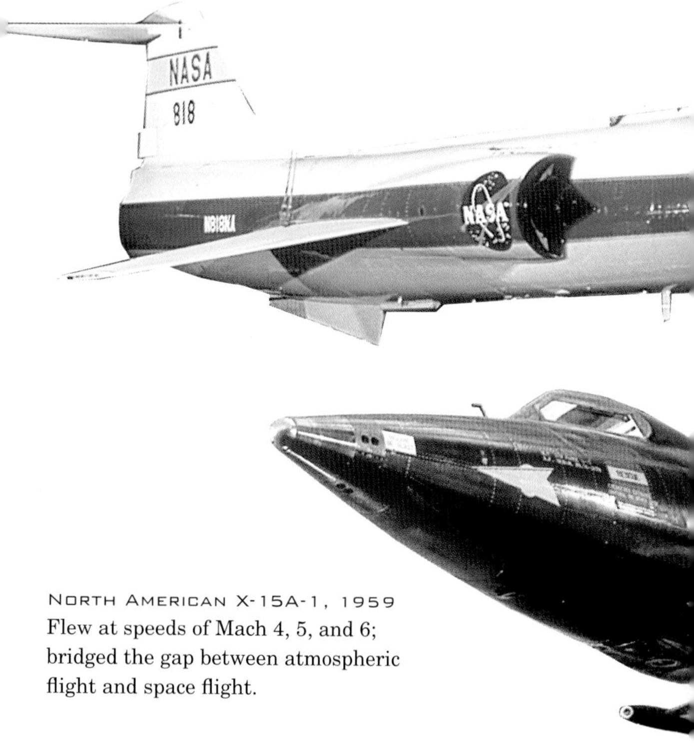

NORTH AMERICAN X-15A-1, 1959
Flew at speeds of Mach 4, 5, and 6;
bridged the gap between atmospheric
flight and space flight.

LOCKHEED F-104
STARFIGHTER, 1958
Developed in 1954, first U.S.
fighter to operate at speeds in
excess of Mach 2.

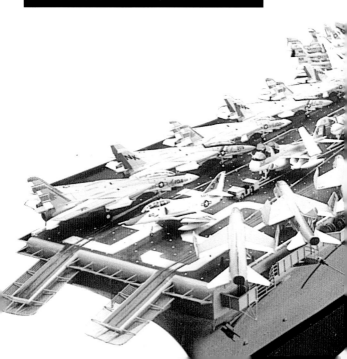

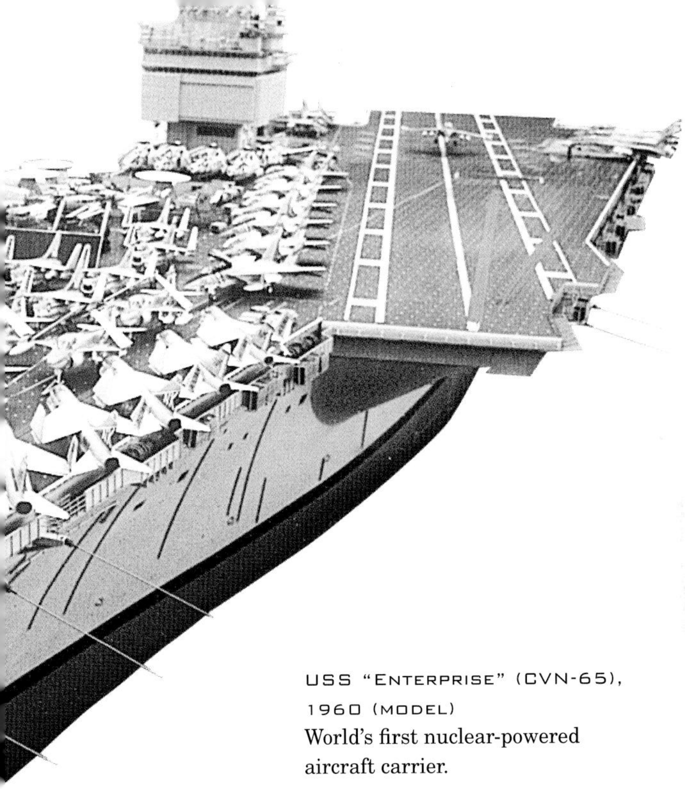

USS "ENTERPRISE" (CVN-65),
1960 (MODEL)
World's first nuclear-powered
aircraft carrier.

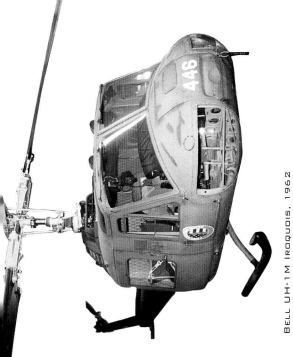

BELL UH-1M IROQUOIS, 1962

A multipurpose helicopter, the "Huey" served in Vietnam as a gunship, troop carrier, and ambulance.

MCDONNELL F-4S PHANTOM II, 1975
The S model is a modified variant of the Phantom, which first flew in 1958 and later served with the Navy, Marines, and Air Force.

163

LOCKHEED SR-71 BLACKBIRD, 1964
High-altitude reconnaissance airplane complemented the U-2;
capable of speeds above Mach 3 and altitudes above eighty
thousand feet.

Hawker Siddeley XV-6A Kestrel, 1968
Forerunner of the Harrier jump jet, its variable-position
nozzles' direct thrust allowing for takeoffs as either
conventional airplane or helicopter.

GATES LEARJET, 1963
First successful business jet.

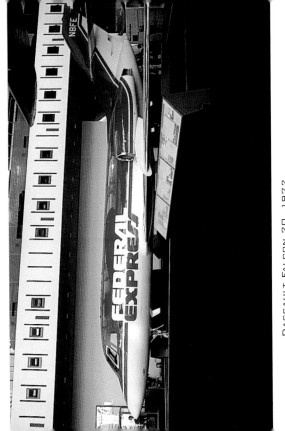

DASSAULT FALCON 20, 1972
A French aircraft, highly successful as a parcel transport.

"DOUBLE EAGLE II" GONDOLA, 1978
The first balloon to fly across the Atlantic, on August 17, 1978, taking 137 hours and 6 minutes.

MacCready Gossamer Condor, 1977
Constructed of lightweight materials; made the first sustained, maneuverable, man-powered flight on August 27, 1977.

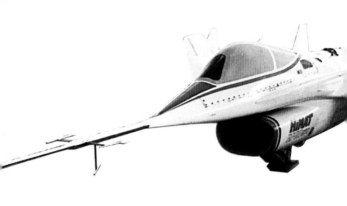

ROCKWELL HiMAT, 1978
Developed to test Highly Maneuverable Aircraft Technology; an
unmanned experimental craft featuring a variable-wing design.

RUTAN VARIEZE, 1978
Designed by Burt Rutan, the VariEze launched a revolution
in home-built aircraft construction technology.

BELL 206L LONG RANGER, 1982
The "SPIRIT OF TEXAS," the first helicopter to fly around the world; completed its month-long flight on September 30, 1982.

CESSNA 150L COMMUTER, 1974
More living pilots have soloed in the Cessna 150 than in any other aircraft.

AMERICAN AEROLIGHTS DOUBLE EAGLE, 1982
First ultralight employed to aid law enforcement by the police department of Monterey Park, California, on September 2, 1982.

GRUMMAN X-29, 1984

First flown on December 14, 1984; its forward-swept wing design tested military applications for highly maneuverable future fighters.

175

Extra 260, 1986
One-of-a-kind aircraft created by Walter Extra. Flown by Patty Wagstaff to win the U.S. National Aerobatic championship in 1991 and 1992.

RUTAN "VOYAGER," 1986
First airplane to circum-
navigate the globe, nonstop,
without refueling.

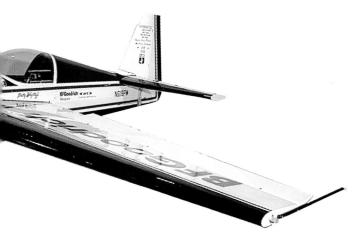

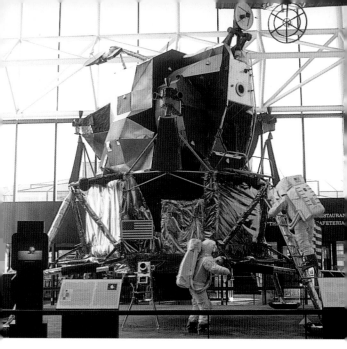

APOLLO LUNAR MODULE (TEST VEHICLE), 1969
Museum's Lunar Lander was intended for rendezvous and
docking tests in Earth orbit.

ROCKETRY AND SPACE FLIGHT

A**N UNDERSTANDING** of the origins of space flight must begin with the earliest rockets. After all, rockets have launched astronauts and their spacecraft into orbit, sent scientific and communications satellites into space, propelled robotic explorers to other worlds in our solar system, and launched men to the Moon. The first rockets, forerunners of modern missiles and space launch vehicles, were probably used as weapons by the Chinese in the twelfth or thirteenth centuries. Rockets were also used in nineteenth-century Europe for military purposes. Modern rocketry laid the foundations for space launch vehicles: Konstantin Tsiolkovsky's theoretical research on rockets and spaceflight in Russia at the turn of the century, Robert Goddard's practical development of liquid-propellant rockets in the United States in the 1920s and 1930s, and the German rocket projects in World War II.

After World War II, the United States and the Soviet Union became embroiled in a struggle for world leadership, and the technologies of rocketry and spaceflight played prominent roles as the two superpowers competed in an arms race and a space race. Successful ventures in space were perceived as symbolic victories of competing social systems, and—since the rocketry that made space-flight possible was also used in the development of ballistic missiles—an achievement in space also represented a step ahead in the international arms race.

In 1961 President John F. Kennedy challenged America to land a man on the Moon before the end of the decade. The Mercury, Gemini, and Apollo programs of the 1960s were steps toward that goal, which was achieved as the world watched on July 20, 1969, when astronauts on the Apollo 11 mission set foot on the Moon. From 1968 through 1972, the United States sent nine manned missions to the Moon, and twelve astronauts landed there. Meanwhile, the Soviet Union launched successful robotic missions to the Moon, but never launched a manned lunar mission.

Throughout this era of much-publicized human space flight, Americans and Soviets also launched hundreds of automated scientific and communications satellites, as well as the first robotic planetary explorers. Both nations also turned their attention to establishing a longer-term human presence in space, the Soviet Union with the Salyut space stations and the United States with Skylab. During 1973 and 1974, the Skylab space station was occupied by three crews who, while in orbit, conducted pioneering solar observations, biomedical investigations, and experiments in materials science. In 1975 the two competing nations met in space in the historic Apollo-Soyuz rendezvous, when American astronauts and Soviet cosmonauts linked their spacecraft as a symbol of a new era of peaceful cooperation.

In the latter half of the 1970s, the United States space program focused on the Viking mission to Mars, the launch of the Voyager spacecraft on a grand tour of the outer planets, and the development of the Space Shuttle, which made its debut in 1981. In the 1980s, the Space Shuttle became the principal United States launch vehicle for deploying and servicing satellites, launching spacecraft on planetary missions, and serving as a laboratory for manned space science research. The Soviets established a new space station, *Mir*, and continued to use their Soyuz spacecraft for trips back and forth to orbit.

As the twentieth century draws to a close, proposals are being debated and work is in progress to establish a permanent international space station. Planetary exploration continues with missions to Venus, Mars, and Jupiter. Astrophysical observatories in Earth orbit probe the universe, seeking new clues for understanding the stars, galaxies, and exotic phenomena such as black holes. As technology continues to advance and more nations take advantage of the commercial and scientific opportunities of spaceflight, the quest to explore the frontiers of space continues.

100 Pr
Rocket
Congreve.
A.D.1815

CONGREVE ROCKET
(MODEL), 1815
Similar to those mentioned
in Francis Scott Key's "Star
Spangled Banner," the first
"modern" ballistic rockets;
utilized sticks for stability.

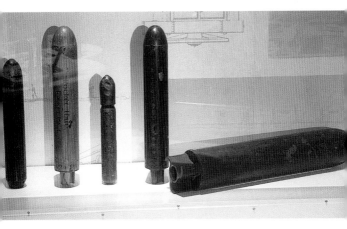

HALE ROCKETS, C. 1850

Spin-stabilized rockets; used by the British during the
Crimean War and by Confederate and Union forces during
the American Civil War.

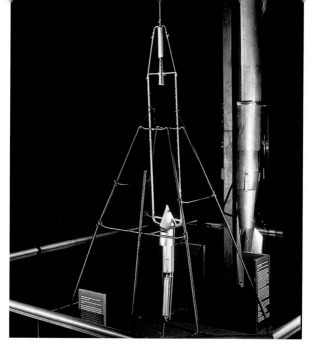

GODDARD 1926 ROCKET (MODEL)

Launched on March 16, 1926; the world's first successfully flown liquid-propellant rocket.

GODDARD 1941 ROCKET

P-series rocket; used gas generator-run centrifugal pumps to introduce propellants into the combustion chamber.

Germany's "Vengeance Weapon 1," the forerunner of modern cruise missiles; operated on a pulse jet engine rather than a rocket motor.

V-2 MISSILE, 1944–45
Germany's "VENGEANCE WEAPON 2," the world's first long-range, liquid-propellant ballistic missile.

HS 298 MISSILE, 1945
German radio-controlled air-to-air guided missile; intended to be launched from an aircraft such as the Dornier Do 217, the Fw 190, or the Junkers Ju 88G; never made operational.

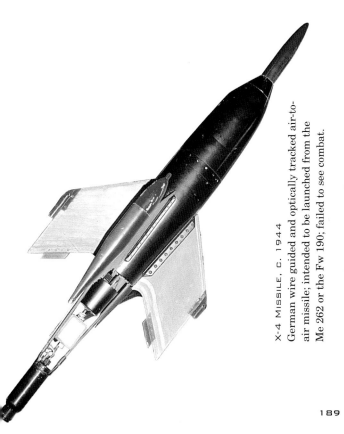

X-4 Missile, c. 1944

German wire guided and optically tracked air-to-air missile; intended to be launched from the Me 262 or the Fw 190; failed to see combat.

RHEINTOCHTER MISSILE, 1944

Germany's two-stage, solid-propellant rocket used as a radio-controlled, surface-to-air missile.

WAC Corporal Rocket, c. 1947

Sounding rocket developed in the late 1940s; in 1949 achieved an altitude record of 250 miles launched as second stage of a V-2 during Project Bumper.

**FURTHER DEVELOPMENTS
IN ROCKETRY AND THE
ADVENT OF THE SATELLITE
(1950–1959)**

VIKING ROCKET, C. 1955 (CENTER)
Developed as a replacement for the aging World War II V-2s
(LEFT) confiscated and used for research in America.

FARSIDE ROCKET, 1957

A research rocket carried aloft by a balloon and launched;
used solid-propellant rocket motors.

JUPITER-C ROCKET, 1958
A modified Redstone ballistic rocket, with three solid-propellant upper stages; launched America's first satellite into orbit.

VANGUARD ROCKET, 1958
A descendant of both the
Viking and Aerobee-Hi
rockets; twelve launches;
placed three satellites
into orbit.

"Sputnik 1" Satellite (replica), 1957
World's first artificial satellite; placed into orbit by the
Soviet Union.

EXPLORER 1 SATELLITE (BACK-UP), 1958
First successful U.S. satellite; launched on January 31, 1958.

Vanguard 1 Satellite
(back-up), 1958–59
Second successful U.S. satellite;
launched on March 17, 1958.

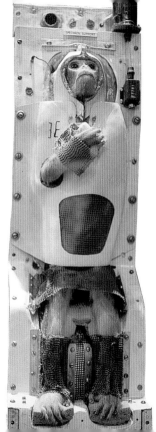

ABLE AND BAKER
BIOCAPSULES, 1959
Two monkeys traveled in
a 300-mile-high ballistic
flight on May 28, 1959.

PIONEER 1 LUNAR PROBE, 1958
First spacecraft to incorporate course-correction rockets and a retro-rocket to allow insertion into a lunar orbit.

PIONEER 4 LUNAR PROBE, 1959
Launched on March 3, 1959; came within 37,300 miles of the Moon.

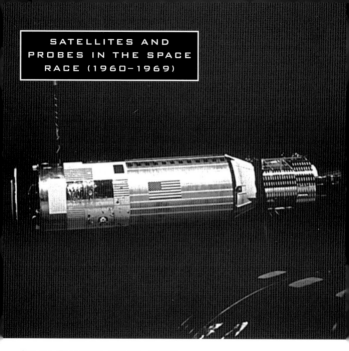

AGENA B UPPER STAGE, 1959

A liquid-propellant upper stage for the Atlas and Thor launch vehicles.

TIROS 1 SATELLITE, 1960
First Television and Infra-red Observation Satellite; launched
on April 1, 1960.

AEROBEE 150, C. 1960
Two-stage rocket; an
improved version of the 1952
Aerobee-Hi model.

AEROBEE 150 NOSE CONE,
c. 1960
From the Naval Research
Laboratory, equipped with
sun-seeker pointing controls.

SOLRAD 1 SATELLITE, 1960
Solar Radiation Satellite examined the effects of solar
radiation on communications and on the Earth's atmosphere;
launched on June 22, 1960.

DISCOVERER 13 SATELLITE, 1960
First man-made object retrieved from space; returned photographic reconnaissance data; launched from a Thor-Agena Rocket on August 10, 1960.

OSO 1 Satellite, 1962

First Orbiting Solar Observatory; studied ultraviolet, x-ray, and gamma radiation from the sun; launched on March 7, 1962.

ARIEL 2 SATELLITE, 1962
One of the first internationally
designed satellites; measured
galactic radiation.

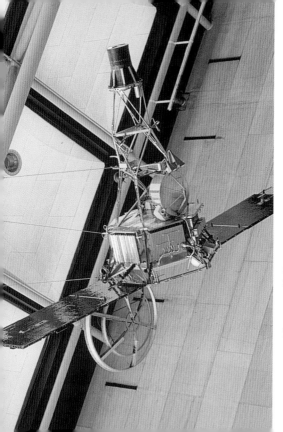

MARINER 2 PLANETARY PROBE (REPLICA), 1962
Directed toward Venus; first spacecraft to fly past
another planet.

"RELAY 1" SATELLITE, 1962
First active communications satellite built under NASA's aegis.

RANGER LUNAR PROBE (REPLICA), 1964
Originally intended as a platform for scientific experiments;
redesigned for lunar mapping.

SURVEYOR LUNAR PROBE (TEST ARTICLE), 1966–68
Landed on the Moon and relayed information about the
Moon's surface in preparation for Apollo lunar landings.

SURVEYOR TELEVISION CAMERA, 1967
Landed on the Moon as part of Surveyor 3 on April 20, 1967; retrieved in 1969 by Apollo 12 crew.

Lunar Orbiter (test article), 1966–67
Photographed, developed, and transmitted scanned pictures of the Moon's surface to Earth in preparation for Apollo lunar landings.

MERCURY SPACECRAFT "FREEDOM 7," 1961
America's first piloted spaceflight on May 5, 1961, with
Alan B. Shepard, Jr., astronaut.

MERCURY SPACECRAFT
"FRIENDSHIP 7," 1962
First American in orbit,
John H. Glenn, Jr., on
February 20, 1962.

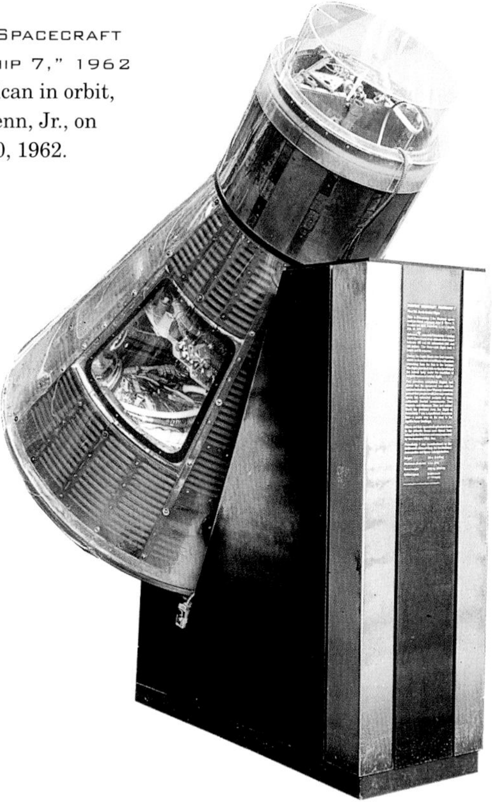

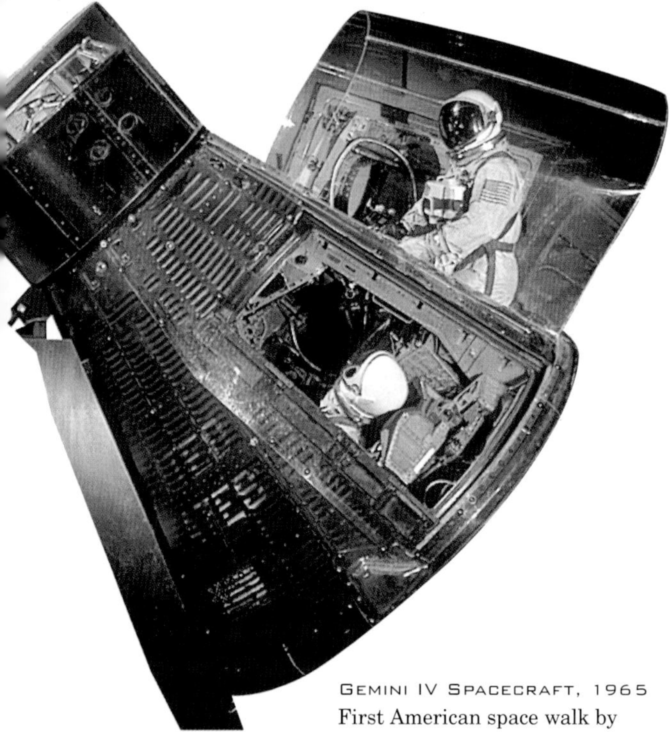

GEMINI IV SPACECRAFT, 1965
First American space walk by
Edward H. White II.

GEMINI VII SPACECRAFT, 1965
Longest U.S. space flight until the 1973 Skylab missions;
Frank Borman and James A. Lovell, Jr., astronauts.

Modified Navy MK IV high-pressure suit, designed as a back-up in case of cabin depressurization.

GEMINI SPACESUIT, 1965–66
Comfortable and mobile, and intended as work clothes as well as a pressurization suit.

APOLLO SPACESUIT, 1967–72
Features a self-contained life-support system; primarily
designed for EVA (Extra-Vehicular Activity).

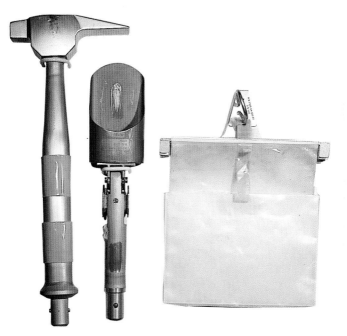

Tools used for collecting
lunar soil and rocks.

APOLLO 11 COMMAND MODULE HATCH, 1969
Could be opened outward in five seconds by pumping the
handle to activate a pressurized nitrogen cylinder.

APOLLO 11 COMMAND MODULE
"COLUMBIA," 1969
Carried Neil Armstrong, Michael
Collins, and Edwin E. "Buzz"
Aldrin into lunar orbit
and back to Earth.

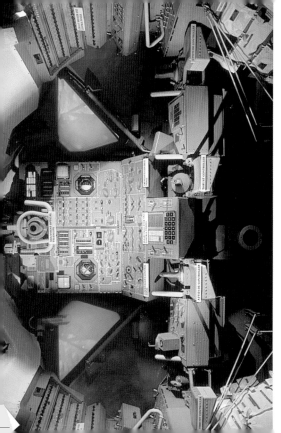

APOLLO LUNAR MODULE COCKPIT (REPLICA), 1969–72
Shows the controls operated by the Apollo 17 astronauts
during their descent to and ascent from the Moon.

LUNAR ROVING VEHICLE (TEST VEHICLE), 1971
Similar to the rovers used on the Moon during the last three Apollo missions.

"UHURU" SATELLITE (SAS-1) (MODEL), 1970
First satellite entirely devoted to the study of x-rays; also first
U.S. satellite launched by a foreign crew.

ITOS 1 Weather Satellite (¹⁄₁₀ scale model), 1970
Second generation TIROS satellite; launched into polar orbit
on January 23, 1970.

PRINCETON EXPERIMENT PACKAGE (FROM OAO), 1972
Launched on August 21, 1972, the largest civilian
space telescope of its time; studied hot stars and the
interstellar medium.

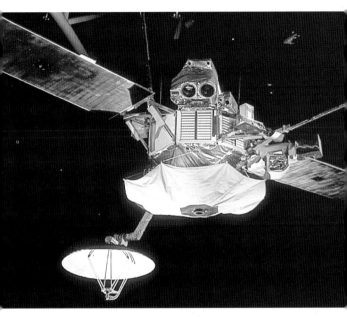

Mariner 10 Planetary Probe (back-up), 1973

First probe to explore two planets (Venus and Mercury) on the same mission.

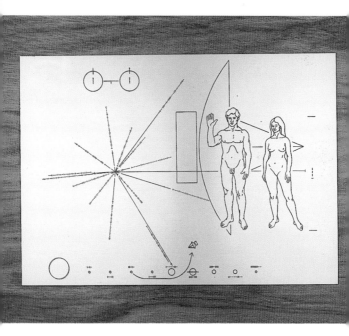

PIONEER PLAQUE, 1972
Same as that attached to the Pioneer 10 and Pioneer 11 planetary
spacecraft; first terrestrial artifact to leave the Solar System.

PIONEER 10 PLANETARY PROBE (PROTOTYPE), 1972
First spacecraft to explore Jupiter; launched on
March 3, 1972.

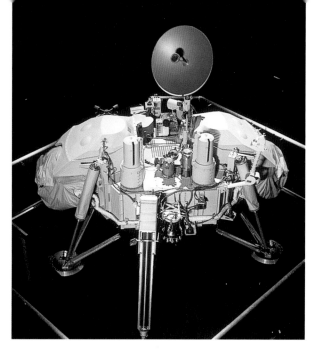

VIKING MARS LANDER (TEST ARTICLE), 1975

Two orbiters and landers launched in 1975; landers touched down in 1976; collected data on Mars until 1980 and 1982.

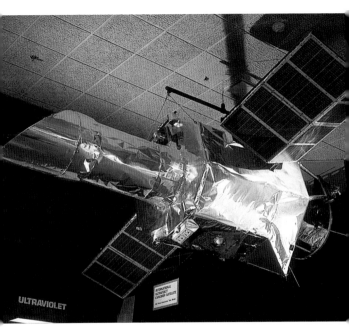

ULTRAVIOLET

IUE Satellite, 1978
International Ultraviolet Explorer Satellite, operating
as an international space observatory; launched on
January 26, 1978.

VOYAGER PLANETARY PROBES, 1977
Provided valuable information about Jupiter, Saturn, Uranus, and Neptune on "grand tour" of the Solar System; launched on August 20 and 26, 1977.

VOYAGER RECORD, 1977
Attached to each of the two Voyager spacecraft now traveling
out of the Solar System; encoded with images, music, and
sounds representative of Earth.

TDRS (MODEL), 1983
Tracking and Data Relay Satellite, deployed by Space
Shuttle; vital link between ground control and spacecraft.

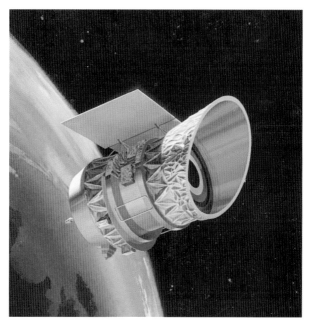

IRAS (MODEL), 1983
Infrared Astronomy Satellite; first unobscured infrared view of the sky.

LANDSAT (¼ SCALE MODEL), 1978
For study of the Earth's natural resources;
first in series launched on March 5, 1978.

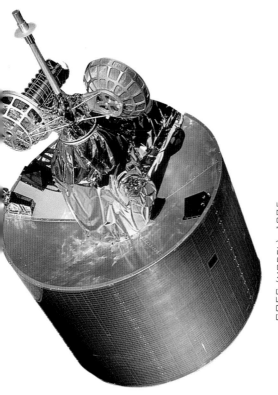

GOES (MODEL), 1975

Geostationary Operational Environmental Satellite; monitors daily weather patterns, severe weather developments, atmospheric temperature, and solar activity.

SKYLAB SPACE STATION (BACK-UP), 1973

Saturn upper stage converted into a habitable workshop;
home and laboratory/observatory for three astronaut crews
in 1973–74.

SKYLAB COMMAND MODULE (BACK-UP), 1973
Same type of command module used in the Apollo
lunar missions.

SKYLAB INTERIOR

Wardroom facilities for food preparation and meals, Earth observation, and crew relaxation.

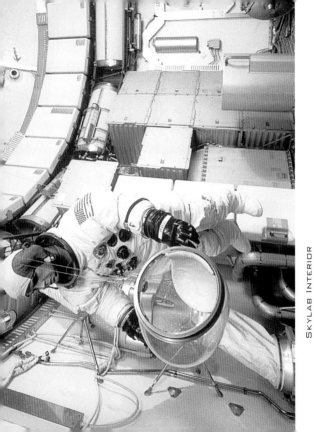

SKYLAB INTERIOR
Experiment and storage area.

SKYLAB APOLLO TELESCOPE MOUNT, 1973
Largest scientific equipment aboard Skylab; featured multiple
telescopes and four, forty-five-foot, electricity-generating
solar panels.

APOLLO-SOYUZ TEST PROJECT, 1975
Apollo 18 (U.S.) and Soyuz 19 (USSR); historic rendezvous in
space on July 17, 1975.

Space Shuttle "Enterprise" (test vehicle), 1977
Primary U.S. launch vehicle in 1980s–1990s; "Enterprise"
used for Space Shuttle landing tests and launch pad tests;
never used in space.

HUBBLE SPACE TELESCOPE
STRUCTURAL DYNAMICS TEST VEHICLE, 1990
Resembles flight observatory in size, basic structure, and dynamic behavior.

HUBBLE SPACE TELESCOPE, 1990 (⅕ SCALE MODEL)
The largest astronomical instrument in space, launched by
the Space Shuttle "DISCOVERY."

SCOUT D ROCKET,
c. 1961 (RIGHT)
Smallest launch vehicle
ever used by NASA;
four-stage, solid-
propellant rocket.

FURTHER DEVELOPMENTS
IN ROCKETS AND BALLISTIC
MISSILES SINCE 1960
(1960–PRESENT)

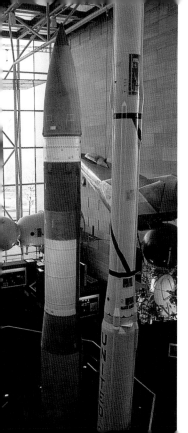

Three-stage, solid-
propellant rocket that
can carry multiple
nuclear warheads.

One of the primary U.S.
intercontinental ballistic
missiles, now banned by
INF treaty.

SS-20 Missile (training version), c. 1975
One of the USSR's primary intercontinental ballistic missiles, capable of carrying three nuclear warheads, now banned by INF treaty.

Saturn V Rocket (¹⁄₄₈ scale model), 1967
First used on November 9, 1967, for the unmanned Apollo 4 flight; retired after launching Skylab into orbit on May 14, 1973; three actual Saturn V rockets in Museum's collection.

Go Baby, Go!

F-1 ROCKET ENGINE, 1967
Cluster of five F-1 rocket engines on first stage of the
Saturn V launch vehicle used for Apollo lunar missions.

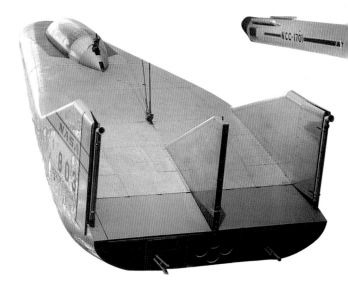

NORTHROP M2-F3 LIFTING BODY, 1967
Experimental aerodynamic lift project, which provided
valuable information for Space Shuttle design.

USS "ENTERPRISE" (MODEL)
Model used in filming original "STAR TREK®" television series, 1966–69.

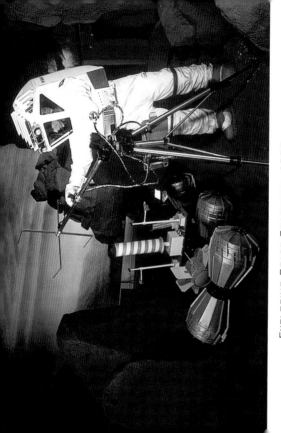

Detail of an advanced spacesuit concept for planetary exploration, displayed beside model of a Russian Mars Rover.

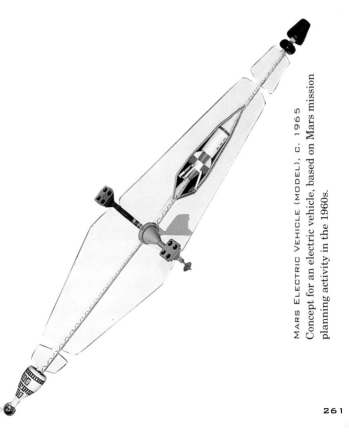

Mars Electric Vehicle (model), c. 1965
Concept for an electric vehicle, based on Mars mission planning activity in the 1960s.

ARCHIVAL AND
ART COLLECTION

THE FIRST THING most visitors see when they enter the National Air and Space Museum is Robert McCall's painting *The Space Mural—A Cosmic View.* Celebrating aviation and spaceflight, McCall's work is part of the art collection the museum has developed along with its other collections, and reflects achievements in astronomy, ballooning, aviation, and spaceflight. It currently includes about thirty-four hundred paintings, drawings, prints, and sculptures, all based on the theme of flight, and is one of the world's primary aeronautical and space-related art repositories.

Some works of art are historical documents in their own right and provide important information that can not be found elsewhere. Many years ago, for example, the museum acquired a collection of 141 prints on ballooning. A curator planning a gallery on ballooning recognized their value as primary source material. Created by artists on the spot, the prints provide an opportunity to see, rather than just imagine, the first successful manned flights. Fifteen hundred "eyewitness" drawings from the National Aeronautics and Space Administration (NASA) and forty-five early aeronautical prints from the American Institute of Aeronautics and Astronautics (AIAA) are other examples of this type of material in the collection. Yet the museum's art collection is much more than a research aid—it also enriches visitors' understanding of the other exhibits in the museum. Artists' unique insights and their diverse artistic perceptions help underscore the historical or sociological message of the shows visitors attend.

The National Air and Space Museum Archives supports the mission of the museum by acquiring and preserving documentary materials relating to air and spaceflight. These include a wide range of visual and textual materials, many emphasizing the technical aspects of air- and spacecraft and their propulsion systems. The archives organizes and describes these materials and assists the public and museum staff in using them in research. The archival collection contains approximately ten thousand cubic feet of material, including some 1.7 million photographs, seven hundred thousand feet of motion picture film, and two million technical drawings.

Although everything in the art collection relates in some way to flight, a wide variety of types and styles of art are represented, from illustrations by Norman Rockwell (page 296) to nonrepresentational sculptures by Morris Graves (page 309). The following pages present a sampling from both the art department and the archives.

Wright, W.

Established in 189_

Langley-Wright controversy

Wright Cycle Company
1127 West Third Street.

DAYTON, OHIO. May 30 1899

The Smithsonian Institution

Washington.

Dear Sirs;

I have been interested in the problem
of mechanical and human flight ever since as a boy
I constructed a number of bats of various sizes after the
style of Cayley's and Penaud's machines. My observations
since have only convinced me more firmly that human flight
is possible and practicable. It is only a question of knowledge
and skill just as in all acrobatic feats. Birds are the
most perfectly trained cars gymnasts in the world and are especially
well fitted for their work, and it may be that man will never
equal them, but no one who has watched a bird chasing an
insect or another bird can doubt that feats are performed which
require three or four times the effort required in ordinary
flight. I believe that simple flight at least is possible to
man and that the _____ experiments and investigations of a large
number of independent workers will result in the accumulation
of information and knowledge and skill which will finally lead to
accomplished flight.

The works on the subject to which I have had access

Wright Cycle Company

1127 West Third Street.

DAYTON, OHIO..................

as Mareys and Jamiesons books published by Appletons and various magazine and cyclopedic articles. I am about to begin a systematic study of the subject in preparation for practical work to which I expect to devote what time I can spare from my regular business. I wish to obtain such papers as the Smithsonian Institution has published on this subject, and if possible a list of other works in print in the English language. I am an enthusiast, but not a crank in the sense that I have some pet theories as to the proper construction of a flying machine. I wish to avail myself of all that is already known and then if possible add my mite to help on the future worker who will attain final success. I do not know the terms on which you send out your publications but if you will inform me of the cost I will remit the price.

Yours truly,
Wilbur Wright.

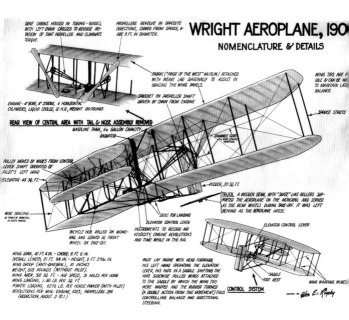

DRIVE CHAINS HOUSED IN TUBING-GUIDES, WITH LEFT CHAIN CROSSED TO REVERSE ROTATION OF THAT PROPELLER AND ELIMINATE TORQUE.

PROPELLERS REVOLVE IN OPPOSITE DIRECTIONS, CARVED FROM SPRUCE, & ARE 8 FT. IN DIAMETER.

WRIGHT AEROPLANE, 190

NOMENCLATURE & DETAILS

FABRIC ("PRIDE OF THE WEST" MUSLIN) ATTACHED WITH WEAVE LAID DIAGONALLY TO ASSIST IN BRACING THE WING PANELS.

WING TIPS ARE F IBLE & CAN BE W TO MAINTAIN LAT BALANCE.

ENGINE- 4" BORE, 4" STROKE, 4 HORIZONTAL CYLINDERS, LIQUID COOLED, 12 H.P., WEIGHT 180 POUNDS

SPROCKET ON PROPELLER SHAFT DRIVEN BY CHAIN FROM ENGINE

SPRUCE STRUTS

REAR VIEW OF CENTRAL AREA WITH TAIL & NOSE ASSEMBLY REMOVED

GASOLINE TANK, ¼ GALLON CAPACITY.
RADIATOR

DYNAMIC GLIDE 4¾° KEEP NEARLY VERTICAL.

PULLEY MOVED BY WIRES FROM CONTROL LEVER SHAFT OPERATED BY PILOT'S LEFT HAND

ELEVATOR 48 SQ. FT.

RUDDER, 20 SQ. FT.

TRUCK. A WOODEN BEAM, WITH "SKATE" LIKE ROLLERS SUPPORTED THE AEROPLANE ON THE MONORAIL AND SERVED AS THE REAR WHEELS DURING TAKE-OFF. IT WAS LEFT BEHIND AS THE AEROPLANE AROSE.

WIRE BRACINGS AT FRONT OF MONORAIL IN OUTER PANELS.

ELEVATOR CONTROL LEVER

SKIDS FOR LANDING

ELEVATOR CONTROL LEVER

BICYCLE HUB ROLLED ON MONORAIL AND SERVED AS FRONT WHEEL ON TAKE-OFF.

INSTRUMENTS TO RECORD AIR VELOCITY, ENGINE REVOLUTIONS AND TIME WHILE IN THE AIR.

WING SPAN, 40 FT. 4 IN. – CHORD, 6 FT. 6 IN.
OVERALL LENGTH, 21 FT. 3⁹/₁₀ IN. – HEIGHT, 9 FT. 3³/₃₂ IN.
WING DROOP (ANTI-DIHEDRAL), 10 INCHES
WEIGHT, 505 POUNDS (WITHOUT PILOT).
WING AREA, 510 SQ. FT. – AIR SPEED, 31 MILES PER HOUR
WING LOADING, 1.46 LB. PER SQ. FT.
POWER LOADING, 62½ LB. PER HORSE POWER (WITH PILOT)
REVOLUTIONS PER MIN: ENGINE, 1025; PROPELLERS 356
(REDUCTION, ABOUT 3 TO 1)

PILOT LAY PRONE WITH HEAD FORWARD, HIS LEFT HAND OPERATING THE ELEVATOR LEVER, HIS HIPS IN A SADDLE. SHIFTING THE HIPS SIDEWISE PULLED WIRES ATTACHED TO THE SADDLE BY WHICH THE WING TIPS WERE WARPED AND THE RUDDER TURNED (A DOUBLE ACTION FROM ONE MOVEMENT) THUS CONTROLLING BALANCE AND DIRECTIONAL STEERING.

SADDLE
FOOT REST

WING WARPING LEVER

CONTROL SYSTEM

DRAWN BY Wm. E. Rigsby

DIAGRAM OF WRIGHT AEROPLANE, 1903

"VIN FIZ" POSTER, C. 1911

After the Wright EX "VIN FIZ" made the first continental
flight across the U.S., Vin Fiz issued this promotional poster.

TRANSPACIFIC

COVER OF ADVERTISING BROCHURE, LATE 1930S
Publicized Pan American Airways transpacific service.

POSTCARD OF TRANSATLANTIC FLIGHT, 1933

The line of stylized Savoia-Marchetti S-55Xs commemorates Italo Balbo's 1933 transatlantic flight.

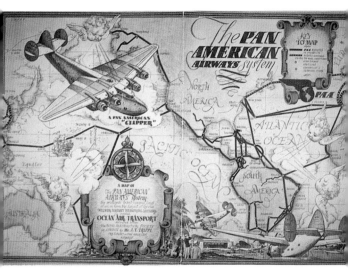

PAN AMERICAN PROGRAM COVER, C. 1941

This map, produced for Juan Trippe's Wilbur Wright Memorial Lecture, shows the extent of the Pan American Airways route system in 1941.

Baggage Label for Pan American Airways, and Affiliate, Mexicana, 1930

BAGGAGE LABEL FOR NATIONAL AIR FORCE (COLOMBIA)
Label reads, "In peace and in war, National Air Force."

PROGRAM COVER FROM CONCURSOS DE AVIACION, 1929
Organized by Real Aero Club de Cataluna as part of 1929
Exposicion Internacional de Barcelona, Spain.

BAGGAGE LABEL FOR THE RUSSIAN AIRLINE DERULUFT
Moscow, Berlin, and Paris are three of the many cities shown
on the map.

CSA is the world's fifth oldest scheduled airline in operation today.

A single-engined airplane over a constellation of stars represents the cities served on Colonial's northeastern U.S. routes.

BAGGAGE LABEL FOR FARMAN AIRLINES
One of the ancestors of Air France, formed in 1919.

"Globe Aerostatique, dédié à M. Charles," late 1700s
Denis, French.
Hand-colored engraving on paper, $10^{3}/_{4} \times 13^{1}/_{2}$ in. (27 × 34 cm).

"MONTGOLFIER FLIGHT," c. 1783
Artist unknown, French.
Tinted engraving on paper, 11 × 8 in. (28 × 20 cm).

"FLYING MEN FROM 'LOS PROVERBIAS,'" 1850 EDITION
Francisco de Goya, Spanish.
Black and white etching on paper, $8\frac{1}{2} \times 13\frac{1}{4}$ in. (22×34 cm).

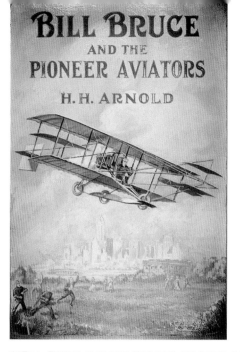

"'Bill Bruce and the Pioneer Aviators,' Early Biplane in Flight," early 1900s

Chris Schaare, American.

Book illustration, oil on millboard, 24¼ × 16½ in. (62 × 42 cm).

"U.S. Air Corps Biplane in Inverted Flight," 1930
Wayne Lambert Davis, American.
Black and white etching on laid paper, 14 × 11 in. (36 × 28 cm).

"Threatening Weather, But the Mail Must Go
Through, 1924," 1983
Wilma Zella Wethington, American.
Watercolor on paper, 22 × 30 in. (56 × 76.2 cm).

"First Commercial Air Mail," 1983
Paul W. Gillan, American.
Oil on canvas, 29½ × 49¼ in. (75 × 126 cm).

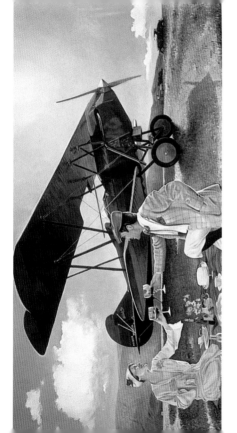

"PICNIC WITH THE TRAVEL AIR," 1983
David Zlotky, American.
Oil on canvas, 23½ × 47½ in. (60 × 121 cm).

"Preparing for Atlantic Survey: Lindbergh's Lockheed Sirius, 1933," 1983

John Paul Jones, American.

Acrylic on canvas, 24 × 48 in. (61 × 122 cm).

"LONE EAGLE," 1983
Kathleen Cantin, American.
Intaglio etching on paper, 8 × 12 in. (20 × 30 cm).

"OVER THE FALLS IN A TRIMOTOR $5.00," 1983
Hugh Laidman, American.
Watercolor on paper, 30 × 36 in. (76.2 × 91 cm).

"The Ultimate Flight," 1983
Lawrence L. Rice, American.
Watercolor on paper, 22 × 30 in. (56 × 76.2 cm).

"P-26 'Pea Shooter,'" 1983
Steven R. Cox, American.
Gouache on illustration board, 30 × 40 in. (76.2 × 101.6 cm).

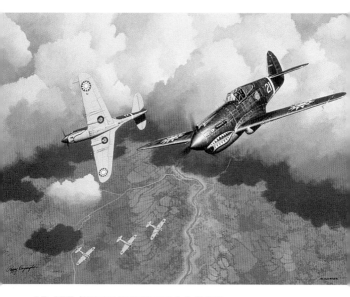

"P-40B (Pappy Boyington)," 1982
Stan Stokes, American.
Acrylic on canvas, 30 × 40 in. (76.2 × 101.6 cm).

"Study for Fortresses Under Fire," 1976
Keith Ferris, American.
Oil on Masonite, 24½ × 74 in. (62 × 188 cm).

"Thunder in the Canyon," 1985
William S. Phillips, American.
Oil on panel, 48 × 96 in. (122 × 244 cm).

"THE EXPLORATION OF MARS," 1953
Chesley Bonestell, American.
Oil on board, 14⅜ × 28 in. (37 × 71 cm).

"Power to Go," 1969
Paul Calle, American.
Oil on panel, $47^{1}/_{2} \times 59^{1}/_{2}$ in. (121 × 151 cm).

"FIRST STEP ON THE MOON," 1966
Norman Rockwell, American.
Oil on canvas, 64 × 40 in. (163 × 101.6 cm).

"SPACE STATION," C. 1977

Jay Mullins, American.

Gouache on paper, 24½ × 34¼ in. (62 × 87 cm).

"THE SPACE MURAL: A COSMIC VIEW
(HORIZONTAL STUDY)," 1975
Robert T. McCall, American.
Acrylic on canvas, 23 × 90 in. (58 × 228.6 cm).

"THE SPACE MURAL:
A COSMIC VIEW
(VERTICAL STUDY),"
1975
Robert T. McCall,
American.
Acrylic on canvas,
$117\frac{1}{2} \times 46$ in.
(298×117 cm).

"K. E. TSIOLKOVSKY," 1975
Anatoliy Yakushin, Russian.
Lithograph on paper, $32\frac{1}{4} \times 22$ in. (82 × 56 cm).

"MAN MADE STARS," 1975
Andrey Konstantinovich Sokolov, Russian.
Oil on Masonite, 24 × 36 in. (61 × 91 cm).

"Highways I," 1976
Ingo Swann, American.
Oil on canvas, 40 × 40 in. (101.6 × 101.6 cm).

"NEW DAY: SKY AND EARTH," 1983
Kathleen Carey, American.
Oil on canvas, 44 × 50 in. (112 × 127 cm).

"CROSSROADS," 1967
Alexander B. Calder, American.
Gouache on paper, $29\frac{3}{8} \times 42\frac{1}{2}$ in. (75 × 108 cm).

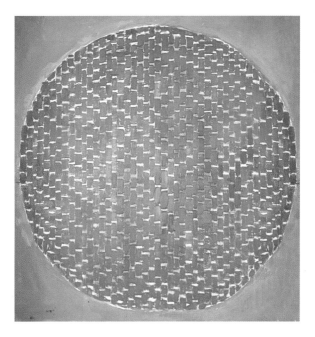

"ASTRONAUTS' GLIMPSE OF THE EARTH," 1974
Alma Woodsey Thomas, American.
Acrylic on canvas, 50 × 50 in. (127 × 127 cm).

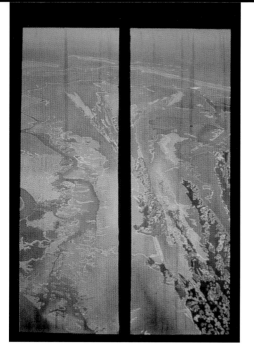

"KIAWAH ISLAND, S.C.," 1985
Mary Edna Fraser, American.
Batik on silk, two panels, 190 × 72 in. (482.6 × 182.9 cm).

"ALITALIA," 1973
Richard Estes, American.
Oil on canvas, 30 × 40 in. (76.2 × 101.6 cm).

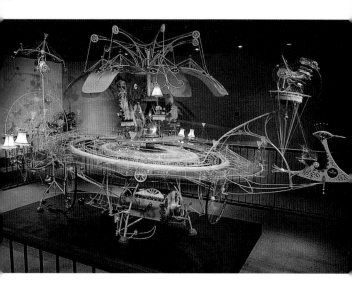

"S.S. Pussiewillow II," 1980
Rowland Emett, O.B.E., British.
Mixed media, including metal, Plexiglas, household objects,
and music, 108 × 216 × 150 in. (274 × 549 × 381 cm).

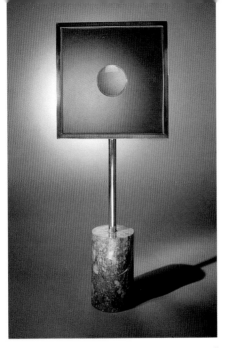

"INSTRUMENTS FOR A NEW NAVIGATION #5," 1961–62
Morris Cole Graves, American.
Glass and chrome-plated steel on marble base,
$25\frac{1}{4} \times 14 \times 1\frac{1}{2}$ in. (64 × 36 × 4 cm).

"Continuum," 1976
Charles Owen Perry, American.
Bronze, 14 ft. diam. (427 cm).

"AD ASTRA," 1976
Richard Lippold, American.
Nickel stainless steel on stainless steel sheets, 110 ft. × 5 ft.
diam. (3353 × 152 cm).

ACKNOWLEDGMENTS

National Air and Space Museum

HEAD OF PUBLICATIONS: Patricia Jamison Graboske
PROJECT COORDINATOR: Helen M. Morrill
RESEARCH INTERNS: John M. Sherrer III, Rupali Gandhi

Special thanks to the following people:

ARCHIVES: Dana Bell, Tim Cronen, Dan Hagedorn,
Kristine Kaske, Melissa Keiser, Brian Nicklas, Tom Soapes

ART DEPARTMENT: Mary Henderson, Susan Lawson-Bell

DEPARTMENT OF AERONAUTICS: Thomas Alison, Dorothy Cochrane,
Tom Crouch, Tom Deitz, Joanne Gernstein, Peter Jakab, Russell Lee,
Michael Neufeld, Dominick Pisano, Robert van der Linden

DEPARTMENT OF SPACE HISTORY: Martin Collins,
David DeVorkin, Valerie Neal, Allan Needell, Frank Winter

PUBLIC AFFAIRS: Michael Fetters

STAFF PHOTOGRAPHERS: Mark Avino, Terry McCrea, Carolyn J. Russo

Abbeville Press

EDITOR: Susan Costello
DESIGNER: Tsang Seymour Design Studio
PRODUCTION EDITOR: Owen Dugan
TYPOGRAPHIC DESIGN: Barbara Sturman
PRODUCTION MANAGER: Lou Bilka

PHOTOGRAPHY CREDITS

Photographs are listed by photographer, when known, then by page number. When available, Smithsonian Institution negative numbers are given in parentheses.

The following photographs are courtesy Smithsonian Institution: 2, 6, 14 (A 26767 B-2), 17 (85-16749), 23, 24 (73-2242), 26, 35 (72-10099), 36–37 (top), 42–43 (top), 46–47 (bottom; 86-6177), 48–49 (bottom), 56, 57 (A 43352), 64 (94-7712), 65 (72-8833), 74 (Beechcraft Collection; 93-15836), 77, 79, 82–83 (bottom), 85, 92–93 (bottom right; 90-1040), 94–95 (top right), 99, 100, 102 (72-8523), 103, 104 (89-1623), 105 (USAF Photo Collection; K 66), 106–107 (bottom), 108, 109 (77-2694), 110 (USAF Photo Collection; K 754), 112–113 (top; USAF Photo Collection; K 3557), 117 (85-16749), 118–119 (top; USAF Photo Collection; K 15609), 118–119 (bottom), 121, 124–125 (bottom; A-43680-E), 126, 127 (72-7871), 128–129 (top; A-29911 A.C.), 128–129 (bottom), 138, 139, 142–143 (top), 146, 150 (A-43306), 152–153 (bottom), 155 (90-14236), 162 (89-5944), 163, 164–165 (bottom; 92-14116), 170–171 (bottom), 171, 172, 174 (84-14598), 187, 190 (91-18725), 192, 199, 203, 205, 207, 209, 210, 214, 219, 222, 228, 233 (79-830B), 236 (80-13717), 247, 254, 257 (87-14654), 264–265, 266 (A 38681), 267 (89-21352), 268 (89-1216), 269 (Museo Aeronautico Caproni; 89-4516), 270 (89-1221), 271 (94-7512), 272 (94-7514), 273 (91-18707), 274 (94-7513), 275 (94-7511), 276 (91-19892), 277 (94-7515), 278, 279, 280, 281, 282, 283, 284, 285 (copyright © 1983 David Zlotky), 286, 287, 288, 289, 290, 291, 292, 293, 294, 295, 296 (Printed by permission of the Norman Rockwell Family Trust. Copyright © 1967 the Norman Rockwell Family Trust.), 297, 298, 299, 300, 301, 302, 303, 304, 305, 306, 307, 308, 310, 311.

Mark Avino (SI-OPPS): 10, 20 (86-14942), 25, 27, 29, 30, 32, 36–37 (bottom; 88-8621), 38, 39, 55 (84-5274), 58 (86-14942), 71 (84-5250), 76, 84, 91 (86-566), 96, 97 (88-10739), 123 (94–8260), 130–131 (bottom; 94-8261), 144–145 (bottom; 94-8263), 151, 157 (86-6050), 167, 176–177 (bottom; 94-7510), 184 (86-14655), 188 (91-13406-26), 189 (94-8272), 196 (87-14645), 217 (87-14656), 223,

232, 237, 241 (87-14648), 251, 259
(STAR TREK © 1994 by Paramount
Pictures. All rights reserved. STAR
TREK is a registered trademark of
Paramount Pictures. Used by per-
mission.), 261.

Richard Farrar (SI-OPPS): 142–143
(bottom; 78-15-1).

Dale Hrabak (SI-OPPS): 34 (79-4639),
40 (82-8329), 46-47 (top; 81-14836),
54 (80-12881), 62 (77-9236), 72,
73, 75, 94–95 (bottom; 83-2943),
101 (84-3515), 116 (83-14510), 122
(79-4623), 148, 149, 164–165 (bottom; 80-13716), 168, 169, 248–249
(85-17423), 309.

Terry McCrea (SI-OPPS): 106–107 (top;
94-8259), 178, 182 (94-8268), 183
(94-8269), 186 (94-8271), 191 (94-
8273), 194 (94-8275), 195 (94-8276),
198 (94-8277), 201 (94-8278), 204,
206 (94-8279), 208 (94-8280), 212
(94-8281), 213 (94-8282), 215 (94-
8283), 229 (94-8286), 230, 235 (94-
8288), 238 (94-8289), 240 (94-8290),
243 (94-8291), 246 (94-8292), 250
(94-8193), 252, 253, 255 (94-8294),
256 (94-8295), 258 (94-8296).

Robert C. Mikesh: 88–89, 173.

Dane Penland (SI-OPPS): cover (79-763),
back cover (79-764), 1 (80-4966), 3
(80-4978), 28 (79-758), 31 (80-2081),
52, 53 (79-831), 59 (79-763), 60 (80-
2095), 61 (80-2101), 63 (80-2104),
66 (80-2092), 67 (80-4972), 68 (80-
2100), 69 (80-2082), 70 (80-2083),
81 (80-2096), 82–83 (top right; 80-

4966), 86 (80-4969), 90 (80-4973),
92–93 (top; 80-4971), 111 (80-2088),
120 (80-2090), 124 (80-2089), 144–
145 (top; 80-4968), 147 (79-762),
152–153 (top; 80-4974), 158–159
(top; 79-760), 158–159 (bottom; 79-
833), 166 (80-4967), 197 (80-4976),
216 (79-757), 218 (80-4977), 225
(80-4978), 226 (79-765), 227 (79-
832), 231 (80-4979), 234 (80-3070),
242 (79-829), 244 (80-2116), 245.

Charles Phillips: 130–131 (top;
86-14947).

Carolyn J. Russo (NASM): 33, 48–49
(top; 94-5761), 78, 80, 87, 112–113
(bottom; 92-6076), 132 (90-14234),
133 (90-14230), 134, 135, 136, 137
(94-8262), 141, 154 (94-8265), 155
(90-14236), 156, 185 (94-8270), 260.

Evan Sheppard: spine, 41, 42–43 (bottom right), 44–45 (bottom right), 50,
51, 114, 115, 160–161, 220, 221, 224.

The following photographs are courtesy National Aeronautics and
Space Administration: 170–171
(top; 94-7713), 175 (92-15101), 193
(76-1705), 200 (73-H-830), 202
(66-H-1024), 211 (62-Relay-17), 239
(271-1423).

Other sources: 44–45 (top; Photo
courtesy of Imperial War Museum,
London, IWM Neg. No. Q58494),
98 (Photo courtesy of Kenneth
Eidnes), 140 (Photograph Courtesy
of Robert C. Mikesh), 176–177 (top;
Printed with permission of Mark
Greenberg/Visions).

INDEX